MATH
WORD PROBLEM

ADDITION

1. 19 red peaches and 15 green peaches are in the basket. How many peaches are in the basket?

2. 25 oranges were in the basket. More oranges were added to the basket. Now there are 45 oranges. How many oranges were added to the basket?

3. Paul has 77 bananas and David has 12 bananas. How many bananas do Paul and David have together?

4. Marcie has 55 more plums than Audrey. Audrey has 45 plums. How many plums does Marcie have?

5. 39 pears were in the basket. 15 are red and the rest are green. How many pears are green?

6. 100 apples were in the basket. More apples were added to the basket. Now there are 120 apples. How many apples were added to the basket?

7. Janet has 56 more balls than Jennifer. Jennifer has 37 balls. How many balls does Janet have?

8. Allan has 65 marbles and Brian has 30 marbles. How many marbles do Allan and Brian have together?

1. Marin has 65 more oranges than Marcie. Marcie has 29 oranges. How many oranges does Marin have?

2. Allan has 40 bananas and Brian has 6 bananas. How many bananas do Allan and Brian have together?

3. 18 red apples and 34 green apples are in the basket. How many apples are in the basket?

4. 44 marbles were in the basket. More marbles were added to the basket. Now there are 94 marbles. How many marbles were added to the basket?

5. 52 balls were in the basket. 6 are red and the rest are green. How many balls are green?

6. Donald has 61 avocados and Steven has 45 avocados. How many avocados do Donald and Steven have together?

7. 84 apricots were in the basket. More apricots were added to the basket. Now there are 134 apricots. How many apricots were added to the basket?

8. 75 red pears and 23 green pears are in the basket. How many pears are in the basket?

1. Brian has 76 apricots and Paul has 49 apricots. How many apricots do Brian and Paul have together?

2. 3 balls were in the basket. More balls were added to the basket. Now there are 34 balls. How many balls were added to the basket?

3. 22 red oranges and 16 green oranges are in the basket. How many oranges are in the basket?

4. Marcie has 29 more avocados than Jennifer. Jennifer has 24 avocados. How many avocados does Marcie have?

5. 107 bananas were in the basket. 96 are red and the rest are green. How many bananas are green?

6. 84 peaches were in the basket. 73 are red and the rest are green. How many peaches are green?

7. Paul has 64 apples and Allan has 14 apples. How many apples do Paul and Allan have together?

8. 34 plums were in the basket. More plums were added to the basket. Now there are 82 plums. How many plums were added to the basket?

1. 34 apricots were in the basket. More apricots were added to the basket. Now there are 35 apricots. How many apricots were added to the basket?

2. 51 red plums and 36 green plums are in the basket. How many plums are in the basket?

3. Janet has 29 more bananas than Marin. Marin has 19 bananas. How many bananas does Janet have?

4. 106 marbles were in the basket. 72 are red and the rest are green. How many marbles are green?

5. Brian has 44 pears and Paul has 36 pears. How many pears do Brian and Paul have together?

6. 11 apples were in the basket. More apples were added to the basket. Now there are 57 apples. How many apples were added to the basket?

7. Brian has 15 oranges and Allan has 26 oranges. How many oranges do Brian and Allan have together?

8. Michele has 63 more balls than Audrey. Audrey has 22 balls. How many balls does Michele have?

1. 35 red avocados and 22 green avocados are in the basket. How many avocados are in the basket?

2. Sandra has 54 more bananas than Jennifer. Jennifer has 16 bananas. How many bananas does Sandra have?

3. David has 38 peaches and Adam has 43 peaches. How many peaches do David and Adam have together?

4. 70 apples were in the basket. More apples were added to the basket. Now there are 83 apples. How many apples were added to the basket?

5. 65 marbles were in the basket. 19 are red and the rest are green. How many marbles are green?

6. 34 pears were in the basket. 26 are red and the rest are green. How many pears are green?

7. 75 red apricots and 50 green apricots are in the basket. How many apricots are in the basket?

8. Adam has 3 balls and Jake has 31 balls. How many balls do Adam and Jake have together?

1. 56 plums were in the basket. 11 are red and the rest are green. How many plums are green?

2. 78 avocados were in the basket. More avocados were added to the basket. Now there are 87 avocados. How many avocados were added to the basket?

3. David has 79 marbles and Adam has 10 marbles. How many marbles do David and Adam have together?

4. 48 red bananas and 27 green bananas are in the basket. How many bananas are in the basket?

5. Sharon has 95 more pears than Sandra. Sandra has 9 pears. How many pears does Sharon have?

6. Ellen has 25 more balls than Michele. Michele has 23 balls. How many balls does Ellen have?

7. Donald has 24 peaches and Adam has 21 peaches. How many peaches do Donald and Adam have together?

8. 33 apples were in the basket. 30 are red and the rest are green. How many apples are green?

1. Jackie has 72 more oranges than Amy. Amy has 50 oranges. How many oranges does Jackie have?

2. 60 apples were in the basket. More apples were added to the basket. Now there are 109 apples. How many apples were added to the basket?

3. Billy has 58 bananas and Paul has 49 bananas. How many bananas do Billy and Paul have together?

4. 28 red avocados and 48 green avocados are in the basket. How many avocados are in the basket?

5. 108 peaches were in the basket. 67 are red and the rest are green. How many peaches are green?

6. 62 apricots were in the basket. 23 are red and the rest are green. How many apricots are green?

7. Jackie has 75 more plums than Marcie. Marcie has 27 plums. How many plums does Jackie have?

8. Adam has 1 ball and Brian has 23 balls. How many balls do Adam and Brian have together?

1. 24 marbles were in the basket. More marbles were added to the basket. Now there are 64 marbles. How many marbles were added to the basket?

2. 57 red plums and 27 green plums are in the basket. How many plums are in the basket?

3. David has 68 bananas and Adam has 7 bananas. How many bananas do David and Adam have together?

4. 109 peaches were in the basket. 87 are red and the rest are green. How many peaches are green?

5. Marin has 4 more oranges than Sharon. Sharon has 47 oranges. How many oranges does Marin have?

6. 51 avocados were in the basket. More avocados were added to the basket. Now there are 63 avocados. How many avocados were added to the basket?

7. Billy has 8 apples and Allan has 39 apples. How many apples do Billy and Allan have together?

8. Marcie has 79 more apricots than Sharon. Sharon has 6 apricots. How many apricots does Marcie have?

1. 55 peaches were in the basket. 52 are red and the rest are green. How many peaches are green?

2. 18 marbles were in the basket. More marbles were added to the basket. Now there are 26 marbles. How many marbles were added to the basket?

3. David has 90 avocados and Billy has 3 avocados. How many avocados do David and Billy have together?

4. Jackie has 66 more pears than Sandra. Sandra has 17 pears. How many pears does Jackie have?

5. 35 red balls and 10 green balls are in the basket. How many balls are in the basket?

6. 116 oranges were in the basket. 93 are red and the rest are green. How many oranges are green?

7. 22 red apricots and 24 green apricots are in the basket. How many apricots are in the basket?

8. Marin has 81 more apples than Amy. Amy has 42 apples. How many apples does Marin have?

1. 123 pears were in the basket. 81 are red and the rest are green. How many pears are green?

2. 6 red apricots and 22 green apricots are in the basket. How many apricots are in the basket?

3. Marcie has 68 more oranges than Ellen. Ellen has 11 oranges. How many oranges does Marcie have?

4. 29 avocados were in the basket. More avocados were added to the basket. Now there are 41 avocados. How many avocados were added to the basket?

5. Steven has 32 peaches and Donald has 35 peaches. How many peaches do Steven and Donald have together?

6. David has 66 apples and Brian has 15 apples. How many apples do David and Brian have together?

7. Janet has 7 more bananas than Jennifer. Jennifer has 14 bananas. How many bananas does Janet have?

8. 111 balls were in the basket. 71 are red and the rest are green. How many balls are green?

SUBTRACTION

1. Some pears were in the basket. 10 pears were taken from the basket. Now there is 1 pear. How many pears were in the basket before some of the pears were taken?

   ....

2. Amy has 33 fewer balls than Janet. Janet has 36 balls. How many balls does Amy have?

   ....

3. 23 avocados are in the basket. 8 avocados are taken out of the basket. How many avocados are in the basket now?

   ....

4. Brian has 6 apples. Steven has 25 apples. How many more apples does Steven have than Brian?

   ....

5. 45 apricots are in the basket. 5 are red and the rest are green. How many apricots are green?

   ....

6. 2 plums were in the basket. Some of the plums were removed from the basket. Now there is 1 plum. How many plums were removed from the basket?

   ....

7. 33 peaches are in the basket. 9 are red and the rest are green. How many peaches are green?

   ....

8. 19 bananas are in the basket. 19 bananas are taken out of the basket. How many bananas are in the basket now?

   ....

1. 49 marbles were in the basket. Some of the marbles were removed from the basket. Now there are 13 marbles. How many marbles were removed from the basket?

2. 4 apples are in the basket. 4 apples are taken out of the basket. How many apples are in the basket now?

3. 49 avocados are in the basket. 47 are red and the rest are green. How many avocados are green?

4. Adam has 9 balls. Donald has 17 balls. How many more balls does Donald have than Adam?

5. Sharon has 1 fewer pear than Jackie. Jackie has 13 pears. How many pears does Sharon have?

6. Some peaches were in the basket. 13 peaches were taken from the basket. Now there are 12 peaches. How many peaches were in the basket before some of the peaches were taken?

7. 10 apricots were in the basket. Some of the apricots were removed from the basket. Now there are 4 apricots. How many apricots were removed from the basket?

8. Janet has 0 fewer bananas than Marin. Marin has 1 banana. How many bananas does Janet have?

1. 27 peaches are in the basket. 1 is red and the rest are green. How many peaches are green?

2. Adam has 4 apples. Brian has 5 apples. How many more apples does Brian have than Adam?

3. Marcie has 10 fewer marbles than Marin. Marin has 15 marbles. How many marbles does Marcie have?

4. Some balls were in the basket. 19 balls were taken from the basket. Now there are 20 balls. How many balls were in the basket before some of the balls were taken?

5. 39 plums are in the basket. 3 plums are taken out of the basket. How many plums are in the basket now?

6. 28 bananas were in the basket. Some of the bananas were removed from the basket. Now there are 5 bananas. How many bananas were removed from the basket?

7. 40 pears are in the basket. 20 are red and the rest are green. How many pears are green?

8. 4 avocados were in the basket. Some of the avocados were removed from the basket. Now there are 3 avocados. How many avocados were removed from the basket?

1. 33 plums are in the basket. 1 is red and the rest are green. How many plums are green?

2. Billy has 1 marble. Jake has 2 marbles. How many more marbles does Jake have than Billy?

3. 25 avocados are in the basket. 11 avocados are taken out of the basket. How many avocados are in the basket now?

4. Some peaches were in the basket. 5 peaches were taken from the basket. Now there is 1 peach. How many peaches were in the basket before some of the peaches were taken?

5. 44 bananas were in the basket. Some of the bananas were removed from the basket. Now there are 31 bananas. How many bananas were removed from the basket?

6. Janet has 0 fewer balls than Janet. Janet has 2 balls. How many balls does Janet have?

7. 3 pears are in the basket. 2 are red and the rest are green. How many pears are green?

8. Jennifer has 15 fewer oranges than Sandra. Sandra has 38 oranges. How many oranges does Jennifer have?

1. Jake has 18 apricots. Brian has 22 apricots. How many more apricots does Brian have than Jake?

2. 8 avocados were in the basket. Some of the avocados were removed from the basket. Now there are 3 avocados. How many avocados were removed from the basket?

3. Marcie has 15 fewer balls than Michele. Michele has 24 balls. How many balls does Marcie have?

4. 19 bananas are in the basket. 18 bananas are taken out of the basket. How many bananas are in the basket now?

5. 38 peaches are in the basket. 13 are red and the rest are green. How many peaches are green?

6. Some marbles were in the basket. 11 marbles were taken from the basket. Now there are 12 marbles. How many marbles were in the basket before some of the marbles were taken?

7. Some plums were in the basket. 9 plums were taken from the basket. Now there are 5 plums. How many plums were in the basket before some of the plums were taken?

8. 3 pears are in the basket. 2 are red and the rest are green. How many pears are green?

1. Marin has 37 fewer marbles than Jennifer. Jennifer has 41 marbles. How many marbles does Marin have?

2. 5 pears are in the basket. 3 are red and the rest are green. How many pears are green?

3. 37 apples were in the basket. Some of the apples were removed from the basket. Now there are 21 apples. How many apples were removed from the basket?

4. Some apricots were in the basket. 11 apricots were taken from the basket. Now there are 0 apricots. How many apricots were in the basket before some of the apricots were taken?

5. 1 peach is in the basket. 1 peach is taken out of the basket. How many peaches are in the basket now?

6. Billy has 1 banana. Adam has 28 bananas. How many more bananas does Adam have than Billy?

7. 50 balls were in the basket. Some of the balls were removed from the basket. Now there are 36 balls. How many balls were removed from the basket?

8. 42 avocados are in the basket. 37 avocados are taken out of the basket. How many avocados are in the basket now?

1. Michele has 1 fewer pear than Marin. Marin has 3 pears. How many pears does Michele have?

2. Billy has 3 oranges. Steven has 25 oranges. How many more oranges does Steven have than Billy?

3. 30 balls were in the basket. Some of the balls were removed from the basket. Now there are 0 balls. How many balls were removed from the basket?

4. Some peaches were in the basket. 26 peaches were taken from the basket. Now there are 7 peaches. How many peaches were in the basket before some of the peaches were taken?

5. 46 marbles are in the basket. 35 marbles are taken out of the basket. How many marbles are in the basket now?

6. 13 apricots are in the basket. 10 are red and the rest are green. How many apricots are green?

7. Michele has 25 fewer avocados than Marin. Marin has 30 avocados. How many avocados does Michele have?

8. 30 apples are in the basket. 7 are red and the rest are green. How many apples are green?

1. Adam has 30 avocados. Brian has 34 avocados. How many more avocados does Brian have than Adam?

2. 35 bananas are in the basket. 22 bananas are taken out of the basket. How many bananas are in the basket now?

3. Some marbles were in the basket. 2 marbles were taken from the basket. Now there are 0 marbles. How many marbles were in the basket before some of the marbles were taken?

4. 17 apricots were in the basket. Some of the apricots were removed from the basket. Now there are 0 apricots. How many apricots were removed from the basket?

5. Michele has 1 fewer peach than Ellen. Ellen has 2 peaches. How many peaches does Michele have?

6. 37 apples are in the basket. 20 are red and the rest are green. How many apples are green?

7. 46 oranges are in the basket. 46 are red and the rest are green. How many oranges are green?

8. 4 pears are in the basket. 3 pears are taken out of the basket. How many pears are in the basket now?

1. Some plums were in the basket. 23 plums were taken from the basket. Now there are 10 plums. How many plums were in the basket before some of the plums were taken?

   ....

2. Marcie has 26 fewer marbles than Michele. Michele has 50 marbles. How many marbles does Marcie have?

   ....

3. 5 apples are in the basket. 5 are red and the rest are green. How many apples are green?

   ....

4. Brian has 25 balls. Jake has 39 balls. How many more balls does Jake have than Brian?

   ....

5. 16 apricots are in the basket. 1 apricot is taken out of the basket. How many apricots are in the basket now?

   ....

6. 44 pears were in the basket. Some of the pears were removed from the basket. Now there are 37 pears. How many pears were removed from the basket?

   ....

7. 10 avocados are in the basket. 6 are red and the rest are green. How many avocados are green?

   ....

8. Some oranges were in the basket. 7 oranges were taken from the basket. Now there are 5 oranges. How many oranges were in the basket before some of the oranges were taken?

   ....

1. 7 balls are in the basket. 1 is red and the rest are green. How many balls are green?

2. Billy has 1 orange. Brian has 2 oranges. How many more oranges does Brian have than Billy?

3. 35 apricots are in the basket. 20 apricots are taken out of the basket. How many apricots are in the basket now?

4. Marcie has 4 fewer marbles than Marin. Marin has 15 marbles. How many marbles does Marcie have?

5. Some bananas were in the basket. 22 bananas were taken from the basket. Now there are 6 bananas. How many bananas were in the basket before some of the bananas were taken?

6. 35 pears were in the basket. Some of the pears were removed from the basket. Now there are 22 pears. How many pears were removed from the basket?

7. 10 peaches are in the basket. 1 is red and the rest are green. How many peaches are green?

8. 37 avocados were in the basket. Some of the avocados were removed from the basket. Now there are 34 avocados. How many avocados were removed from the basket?

MULTIPLICATION

1. Jackie's garden has 10 rows of pumpkins. Each row has nine pumpkins. How many pumpkins does Jackie have in all?

   ....

2. Paul has two times more balls than Steven. Steven has five balls. How many balls does Paul have?

   ....

3. Allan can cycle eight miles per hour. How far can Allan cycle in one hours?

   ....

4. If there is one apricot in each box and there is one boxes, how many apricots are there in total?

   ....

5. Audrey swims three laps every day. How many laps will Audrey swim in six days?

   ....

6. David can cycle three miles per hour. How far can David cycle in two hours?

   ....

7. Sharon's garden has 14 rows of pumpkins. Each row has 12 pumpkins. How many pumpkins does Sharon have in all?

   ....

8. If there are 11 marbles in each box and there are two boxes, how many marbles are there in total?

   ....

1. Marcie has seven times more avocados than Adam. Adam has two avocados. How many avocados does Marcie have?

2. If there are 13 bananas in each box and there are two boxes, how many bananas are there in total?

3. Jake can cycle three miles per hour. How far can Jake cycle in 11 hours?

4. Jake swims six laps every day. How many laps will Jake swim in 13 days?

5. Michele's garden has nine rows of pumpkins. Each row has 11 pumpkins. How many pumpkins does Michele have in all?

6. Audrey's garden has 14 rows of pumpkins. Each row has four pumpkins. How many pumpkins does Audrey have in all?

7. Allan can cycle nine miles per hour. How far can Allan cycle in six hours?

8. If there are two pears in each box and there are 13 boxes, how many pears are there in total?

1. Billy swims 12 laps every day. How many laps will Billy swim in 11 days?

2. David can cycle seven miles per hour. How far can David cycle in six hours?

3. Janet's garden has 12 rows of pumpkins. Each row has five pumpkins. How many pumpkins does Janet have in all?

4. Paul has 15 times more plums than Janet. Janet has 15 plums. How many plums does Paul have?

5. If there are four avocados in each box and there are six boxes, how many avocados are there in total?

6. Steven can cycle nine miles per hour. How far can Steven cycle in nine hours?

7. Brian has seven times more apricots than Jake. Jake has six apricots. How many apricots does Brian have?

8. Audrey swims three laps every day. How many laps will Audrey swim in 15 days?

1. Amy's garden has three rows of pumpkins. Each row has five pumpkins. How many pumpkins does Amy have in all?

2. Jake swims two laps every day. How many laps will Jake swim in 15 days?

3. If there are four apples in each box and there are 13 boxes, how many apples are there in total?

4. Allan can cycle seven miles per hour. How far can Allan cycle in seven hours?

5. Janet has eight times more pears than Michele. Michele has nine pears. How many pears does Janet have?

6. Jackie's garden has five rows of pumpkins. Each row has nine pumpkins. How many pumpkins does Jackie have in all?

7. Adam can cycle two miles per hour. How far can Adam cycle in nine hours?

8. Marcie swims 14 laps every day. How many laps will Marcie swim in nine days?

1. Brian can cycle 13 miles per hour. How far can Brian cycle in 13 hours?

2. Audrey's garden has seven rows of pumpkins. Each row has 11 pumpkins. How many pumpkins does Audrey have in all?

3. If there are three marbles in each box and there are nine boxes, how many marbles are there in total?

4. Ellen swims six laps every day. How many laps will Ellen swim in three days?

5. Sandra has one times more balls than Jackie. Jackie has 14 balls. How many balls does Sandra have?

6. If there are 10 peaches in each box and there are eight boxes, how many peaches are there in total?

7. Adam swims three laps every day. How many laps will Adam swim in seven days?

8. Marin has 10 times more oranges than Allan. Allan has nine oranges. How many oranges does Marin have?

1. Billy can cycle 15 miles per hour. How far can Billy cycle in nine hours?

2. Marcie's garden has 10 rows of pumpkins. Each row has one pumpkins. How many pumpkins does Marcie have in all?

3. Jackie has three times more apples than Adam. Adam has one apple. How many apples does Jackie have?

4. If there are six peaches in each box and there are 15 boxes, how many peaches are there in total?

5. Audrey swims eight laps every day. How many laps will Audrey swim in eight days?

6. Marcie has 14 times more balls than Janet. Janet has six balls. How many balls does Marcie have?

7. Paul swims nine laps every day. How many laps will Paul swim in seven days?

8. Janet's garden has 10 rows of pumpkins. Each row has four pumpkins. How many pumpkins does Janet have in all?

1. Billy swims 14 laps every day. How many laps will Billy swim in 10 days?

2. If there are seven oranges in each box and there are five boxes, how many oranges are there in total?

3. Sandra has two times more balls than Steven. Steven has 13 balls. How many balls does Sandra have?

4. Janet's garden has two rows of pumpkins. Each row has nine pumpkins. How many pumpkins does Janet have in all?

5. Brian can cycle 11 miles per hour. How far can Brian cycle in 11 hours?

6. Jennifer has 10 times more pears than Sharon. Sharon has 13 pears. How many pears does Jennifer have?

7. David swims two laps every day. How many laps will David swim in two days?

8. Allan can cycle nine miles per hour. How far can Allan cycle in 12 hours?

Page 28

1. Allan swims four laps every day. How many laps will Allan swim in 10 days?

2. Amy's garden has two rows of pumpkins. Each row has six pumpkins. How many pumpkins does Amy have in all?

3. Sharon has 12 times more oranges than Audrey. Audrey has 13 oranges. How many oranges does Sharon have?

4. Paul can cycle one miles per hour. How far can Paul cycle in nine hours?

5. If there are 14 marbles in each box and there are two boxes, how many marbles are there in total?

6. David has one times more pears than Marin. Marin has 11 pears. How many pears does David have?

7. If there are three apples in each box and there are seven boxes, how many apples are there in total?

8. Sandra swims five laps every day. How many laps will Sandra swim in 12 days?

1. Jackie has 13 times more bananas than Janet. Janet has two bananas. How many bananas does Jackie have?

2. Jackie's garden has 10 rows of pumpkins. Each row has 12 pumpkins. How many pumpkins does Jackie have in all?

3. Donald swims 13 laps every day. How many laps will Donald swim in 12 days?

4. David can cycle one miles per hour. How far can David cycle in 11 hours?

5. If there are eight apricots in each box and there are 13 boxes, how many apricots are there in total?

6. Allan can cycle 13 miles per hour. How far can Allan cycle in four hours?

7. If there are six oranges in each box and there are four boxes, how many oranges are there in total?

8. Brian swims six laps every day. How many laps will Brian swim in four days?

1. Allan has five times more apricots than Marin. Marin has seven apricots. How many apricots does Allan have?

2. David can cycle three miles per hour. How far can David cycle in 11 hours?

3. Marin's garden has one rows of pumpkins. Each row has three pumpkins. How many pumpkins does Marin have in all?

4. If there are nine apples in each box and there are 14 boxes, how many apples are there in total?

5. Jake swims 12 laps every day. How many laps will Jake swim in 15 days?

6. Janet has 11 times more marbles than Marcie. Marcie has one marble. How many marbles does Janet have?

7. If there are 12 oranges in each box and there are seven boxes, how many oranges are there in total?

8. Sandra's garden has five rows of pumpkins. Each row has 11 pumpkins. How many pumpkins does Sandra have in all?

DIVISION

1. Billy ordered 12 pizzas. The bill for the pizzas came to $132. What was the cost of each pizza?

2. Jake is reading a book with 57 pages. If Jake wants to read the same number of pages every day, how many pages would Jake have to read each day to finish in 19 days?

3. A box of avocados weighs 266 pounds. If one avocados weighs 19 pounds, how many avocados are there in the box?

4. How many 2 cm pieces of rope can you cut from a rope that is 22 cm long?

5. You have 117 apricots and want to share them equally with 9 people. How many apricots would each person get?

6. Marcie made 323 cookies for a bake sale. She put the cookies in bags, with 19 cookies in each bag. How many bags did she have for the bake sale?

7. You have 104 apples and want to share them equally with 8 people. How many apples would each person get?

8. Jackie made 36 cookies for a bake sale. She put the cookies in bags, with 2 cookies in each bag. How many bags did she have for the bake sale?

1. A box of apricots weighs 112 pounds. If one apricots weighs 7 pounds, how many apricots are there in the box?

2. Brian is reading a book with 44 pages. If Brian wants to read the same number of pages every day, how many pages would Brian have to read each day to finish in 11 days?

3. Marin made 36 cookies for a bake sale. She put the cookies in bags, with 2 cookies in each bag. How many bags did she have for the bake sale?

4. You have 9 marbles and want to share them equally with 9 people. How many marbles would each person get?

5. Steven ordered 3 pizzas. The bill for the pizzas came to $24. What was the cost of each pizza?

6. How many 12 cm pieces of rope can you cut from a rope that is 156 cm long?

7. Marcie made 90 cookies for a bake sale. She put the cookies in bags, with 9 cookies in each bag. How many bags did she have for the bake sale?

8. You have 238 avocados and want to share them equally with 14 people. How many avocados would each person get?

1. Sandra made 28 cookies for a bake sale. She put the cookies in bags, with 7 cookies in each bag. How many bags did she have for the bake sale?

2. How many 14 cm pieces of rope can you cut from a rope that is 14 cm long?

3. Adam is reading a book with 176 pages. If Adam wants to read the same number of pages every day, how many pages would Adam have to read each day to finish in 16 days?

4. Allan ordered 16 pizzas. The bill for the pizzas came to $304. What was the cost of each pizza?

5. A box of peaches weighs 38 pounds. If one peaches weighs 2 pounds, how many peaches are there in the box?

6. You have 42 plums and want to share them equally with 7 people. How many plums would each person get?

7. Sharon made 190 cookies for a bake sale. She put the cookies in bags, with 19 cookies in each bag. How many bags did she have for the bake sale?

8. A box of apricots weighs 20 pounds. If one apricots weighs 4 pounds, how many apricots are there in the box?

1. Donald ordered 16 pizzas. The bill for the pizzas came to $304. What was the cost of each pizza?

2. How many 12 cm pieces of rope can you cut from a rope that is 144 cm long?

3. Billy is reading a book with 88 pages. If Billy wants to read the same number of pages every day, how many pages would Billy have to read each day to finish in 11 days?

4. You have 88 pears and want to share them equally with 8 people. How many pears would each person get?

5. Amy made 95 cookies for a bake sale. She put the cookies in bags, with 5 cookies in each bag. How many bags did she have for the bake sale?

6. A box of plums weighs 10 pounds. If one plums weighs 5 pounds, how many plums are there in the box?

7. How many 3 cm pieces of rope can you cut from a rope that is 48 cm long?

8. Billy is reading a book with 240 pages. If Billy wants to read the same number of pages every day, how many pages would Billy have to read each day to finish in 15 days?

1. You have 132 apples and want to share them equally with 12 people. How many apples would each person get?

2. How many 12 cm pieces of rope can you cut from a rope that is 180 cm long?

3. A box of apricots weighs 40 pounds. If one apricots weighs 20 pounds, how many apricots are there in the box?

4. Sharon ordered 7 pizzas. The bill for the pizzas came to $56. What was the cost of each pizza?

5. Jake is reading a book with 195 pages. If Jake wants to read the same number of pages every day, how many pages would Jake have to read each day to finish in 15 days?

6. Amy made 48 cookies for a bake sale. She put the cookies in bags, with 8 cookies in each bag. How many bags did she have for the bake sale?

7. Marcie ordered 12 pizzas. The bill for the pizzas came to $96. What was the cost of each pizza?

8. A box of avocados weighs 266 pounds. If one avocados weighs 19 pounds, how many avocados are there in the box?

1. Jennifer made 24 cookies for a bake sale. She put the cookies in bags, with 4 cookies in each bag. How many bags did she have for the bake sale?

2. How many 5 cm pieces of rope can you cut from a rope that is 95 cm long?

3. You have 165 plums and want to share them equally with 15 people. How many plums would each person get?

4. Billy is reading a book with 21 pages. If Billy wants to read the same number of pages every day, how many pages would Billy have to read each day to finish in 3 days?

5. A box of oranges weighs 42 pounds. If one oranges weighs 3 pounds, how many oranges are there in the box?

6. Audrey ordered 16 pizzas. The bill for the pizzas came to $112. What was the cost of each pizza?

7. Allan is reading a book with 150 pages. If Allan wants to read the same number of pages every day, how many pages would Allan have to read each day to finish in 10 days?

8. Allan ordered 15 pizzas. The bill for the pizzas came to $240. What was the cost of each pizza?

1. David ordered 9 pizzas. The bill for the pizzas came to $45. What was the cost of each pizza?

2. Donald is reading a book with 30 pages. If Donald wants to read the same number of pages every day, how many pages would Donald have to read each day to finish in 2 days?

3. A box of apricots weighs 32 pounds. If one apricots weighs 4 pounds, how many apricots are there in the box?

4. You have 360 pears and want to share them equally with 18 people. How many pears would each person get?

5. Amy made 126 cookies for a bake sale. She put the cookies in bags, with 14 cookies in each bag. How many bags did she have for the bake sale?

6. How many 7 cm pieces of rope can you cut from a rope that is 35 cm long?

7. Brian ordered 18 pizzas. The bill for the pizzas came to $288. What was the cost of each pizza?

8. How many 4 cm pieces of rope can you cut from a rope that is 68 cm long?

1. You have 84 apricots and want to share them equally with 6 people. How many apricots would each person get?

2. How many 14 cm pieces of rope can you cut from a rope that is 42 cm long?

3. Sandra made 60 cookies for a bake sale. She put the cookies in bags, with 20 cookies in each bag. How many bags did she have for the bake sale?

4. Jennifer ordered 9 pizzas. The bill for the pizzas came to $27. What was the cost of each pizza?

5. A box of marbles weighs 20 pounds. If one marbles weighs 1 pounds, how many marbles are there in the box?

6. Adam is reading a book with 11 pages. If Adam wants to read the same number of pages every day, how many pages would Adam have to read each day to finish in 1 days?

7. Adam ordered 19 pizzas. The bill for the pizzas came to $19. What was the cost of each pizza?

8. A box of bananas weighs 100 pounds. If one bananas weighs 5 pounds, how many bananas are there in the box?

1. Billy is reading a book with 160 pages. If Billy wants to read the same number of pages every day, how many pages would Billy have to read each day to finish in 8 days?

   ....

2. Donald ordered 18 pizzas. The bill for the pizzas came to $144. What was the cost of each pizza?

   ....

3. A box of apricots weighs 12 pounds. If one apricots weighs 4 pounds, how many apricots are there in the box?

   ....

4. You have 136 marbles and want to share them equally with 8 people. How many marbles would each person get?

   ....

5. Marcie made 64 cookies for a bake sale. She put the cookies in bags, with 8 cookies in each bag. How many bags did she have for the bake sale?

   ....

6. How many 20 cm pieces of rope can you cut from a rope that is 20 cm long?

   ....

7. A box of balls weighs 221 pounds. If one balls weighs 13 pounds, how many balls are there in the box?

   ....

8. You have 200 pears and want to share them equally with 20 people. How many pears would each person get?

   ....

1. Janet ordered 12 pizzas. The bill for the pizzas came to $36. What was the cost of each pizza?

2. How many 12 cm pieces of rope can you cut from a rope that is 156 cm long?

3. Brian is reading a book with 8 pages. If Brian wants to read the same number of pages every day, how many pages would Brian have to read each day to finish in 1 days?

4. You have 72 marbles and want to share them equally with 4 people. How many marbles would each person get?

5. A box of pears weighs 180 pounds. If one pears weighs 10 pounds, how many pears are there in the box?

6. Amy made 165 cookies for a bake sale. She put the cookies in bags, with 15 cookies in each bag. How many bags did she have for the bake sale?

7. You have 40 plums and want to share them equally with 4 people. How many plums would each person get?

8. Jackie made 24 cookies for a bake sale. She put the cookies in bags, with 6 cookies in each bag. How many bags did she have for the bake sale?

ANSWERS

Page 1

1. 19 red peaches and 15 green peaches are in the basket. How many peaches are in the basket?
   34

2. 25 oranges were in the basket. More oranges were added to the basket. Now there are 45 oranges. How many oranges were added to the basket?
   20

3. Paul has 77 bananas and David has 12 bananas. How many bananas do Paul and David have together?
   89

4. Marcie has 55 more plums than Audrey. Audrey has 45 plums. How many plums does Marcie have?
   100

5. 39 pears were in the basket. 15 are red and the rest are green. How many pears are green?
   24

6. 100 apples were in the basket. More apples were added to the basket. Now there are 120 apples. How many apples were added to the basket?
   20

7. Janet has 56 more balls than Jennifer. Jennifer has 37 balls. How many balls does Janet have?
   93

8. Allan has 65 marbles and Brian has 30 marbles. How many marbles do Allan and Brian have together?
   95

Page 2

1. Marin has 65 more oranges than Marcie. Marcie has 29 oranges. How many oranges does Marin have?
   94

2. Allan has 40 bananas and Brian has 6 bananas. How many bananas do Allan and Brian have together?
   46

3. 18 red apples and 34 green apples are in the basket. How many apples are in the basket?
   52

4. 44 marbles were in the basket. More marbles were added to the basket. Now there are 94 marbles. How many marbles were added to the basket?
   50

5. 52 balls were in the basket. 6 are red and the rest are green. How many balls are green?
   46

6. Donald has 61 avocados and Steven has 45 avocados. How many avocados do Donald and Steven have together?
   106

7. 84 apricots were in the basket. More apricots were added to the basket. Now there are 134 apricots. How many apricots were added to the basket?
   50

8. 75 red pears and 23 green pears are in the basket. How many pears are in the basket?
   98

Page 3

1. Brian has 76 apricots and Paul has 49 apricots. How many apricots do Brian and Paul have together?
   125

2. 3 balls were in the basket. More balls were added to the basket. Now there are 34 balls. How many balls were added to the basket?
   31

3. 22 red oranges and 16 green oranges are in the basket. How many oranges are in the basket?
   38

4. Marcie has 29 more avocados than Jennifer. Jennifer has 24 avocados. How many avocados does Marcie have?
   53

5. 107 bananas were in the basket. 96 are red and the rest are green. How many bananas are green?
   11

6. 84 peaches were in the basket. 73 are red and the rest are green. How many peaches are green?
   11

7. Paul has 64 apples and Allan has 14 apples. How many apples do Paul and Allan have together?
   78

8. 34 plums were in the basket. More plums were added to the basket. Now there are 82 plums. How many plums were added to the basket?
   48

Page 4

1. 34 apricots were in the basket. More apricots were added to the basket. Now there are 35 apricots. How many apricots were added to the basket?
   1

2. 51 red plums and 36 green plums are in the basket. How many plums are in the basket?
   87

3. Janet has 29 more bananas than Marin. Marin has 19 bananas. How many bananas does Janet have?
   48

4. 106 marbles were in the basket. 72 are red and the rest are green. How many marbles are green?
   34

5. Brian has 44 pears and Paul has 36 pears. How many pears do Brian and Paul have together?
   80

6. 11 apples were in the basket. More apples were added to the basket. Now there are 57 apples. How many apples were added to the basket?
   46

7. Brian has 15 oranges and Allan has 26 oranges. How many oranges do Brian and Allan have together?
   41

8. Michele has 63 more balls than Audrey. Audrey has 22 balls. How many balls does Michele have?
   85

Page 5

1. 35 red avocados and 22 green avocados are in the basket. How many avocados are in the basket?
   57

2. Sandra has 54 more bananas than Jennifer. Jennifer has 16 bananas. How many bananas does Sandra have?
   70

3. David has 38 peaches and Adam has 43 peaches. How many peaches do David and Adam have together?
   81

4. 70 apples were in the basket. More apples were added to the basket. Now there are 83 apples. How many apples were added to the basket?
   13

5. 65 marbles were in the basket. 19 are red and the rest are green. How many marbles are green?
   46

6. 34 pears were in the basket. 26 are red and the rest are green. How many pears are green?
   8

7. 75 red apricots and 50 green apricots are in the basket. How many apricots are in the basket?
   125

8. Adam has 3 balls and Jake has 31 balls. How many balls do Adam and Jake have together?
   34

Page 6

1. 56 plums were in the basket. 11 are red and the rest are green. How many plums are green?
   45

2. 78 avocados were in the basket. More avocados were added to the basket. Now there are 87 avocados. How many avocados were added to the basket?
   9

3. David has 79 marbles and Adam has 10 marbles. How many marbles do David and Adam have together?
   89

4. 48 red bananas and 27 green bananas are in the basket. How many bananas are in the basket?
   75

5. Sharon has 95 more pears than Sandra. Sandra has 9 pears. How many pears does Sharon have?
   104

6. Ellen has 25 more balls than Michele. Michele has 23 balls. How many balls does Ellen have?
   48

7. Donald has 24 peaches and Adam has 21 peaches. How many peaches do Donald and Adam have together?
   45

8. 33 apples were in the basket. 30 are red and the rest are green. How many apples are green?
   3

Page 7

1. Jackie has 72 more oranges than Amy. Amy has 50 oranges. How many oranges does Jackie have?
   122

2. 60 apples were in the basket. More apples were added to the basket. Now there are 109 apples. How many apples were added to the basket?
   49

3. Billy has 58 bananas and Paul has 49 bananas. How many bananas do Billy and Paul have together?
   107

4. 28 red avocados and 48 green avocados are in the basket. How many avocados are in the basket?
   76

5. 108 peaches were in the basket. 67 are red and the rest are green. How many peaches are green?
   41

6. 62 apricots were in the basket. 23 are red and the rest are green. How many apricots are green?
   39

7. Jackie has 75 more plums than Marcie. Marcie has 27 plums. How many plums does Jackie have?
   102

8. Adam has 1 ball and Brian has 23 balls. How many balls do Adam and Brian have together?
   24

Page 8

1. 24 marbles were in the basket. More marbles were added to the basket. Now there are 64 marbles. How many marbles were added to the basket?
   40

2. 57 red plums and 27 green plums are in the basket. How many plums are in the basket?
   84

3. David has 68 bananas and Adam has 7 bananas. How many bananas do David and Adam have together?
   75

4. 109 peaches were in the basket. 87 are red and the rest are green. How many peaches are green?
   22

5. Marin has 4 more oranges than Sharon. Sharon has 47 oranges. How many oranges does Marin have?
   51

6. 51 avocados were in the basket. More avocados were added to the basket. Now there are 63 avocados. How many avocados were added to the basket?
   12

7. Billy has 8 apples and Allan has 39 apples. How many apples do Billy and Allan have together?
   47

8. Marcie has 79 more apricots than Sharon. Sharon has 6 apricots. How many apricots does Marcie have?
   85

Page 9

1. 55 peaches were in the basket. 52 are red and the rest are green. How many peaches are green?
   3

2. 18 marbles were in the basket. More marbles were added to the basket. Now there are 26 marbles. How many marbles were added to the basket?
   8

3. David has 90 avocados and Billy has 3 avocados. How many avocados do David and Billy have together?
   93

4. Jackie has 66 more pears than Sandra. Sandra has 17 pears. How many pears does Jackie have?
   83

5. 35 red balls and 10 green balls are in the basket. How many balls are in the basket?
   45

6. 116 oranges were in the basket. 93 are red and the rest are green. How many oranges are green?
   23

7. 22 red apricots and 24 green apricots are in the basket. How many apricots are in the basket?
   46

8. Marin has 81 more apples than Amy. Amy has 42 apples. How many apples does Marin have?
   123

Page 10

1. 123 pears were in the basket. 81 are red and the rest are green. How many pears are green?
   42

2. 6 red apricots and 22 green apricots are in the basket. How many apricots are in the basket?
   28

3. Marcie has 68 more oranges than Ellen. Ellen has 11 oranges. How many oranges does Marcie have?
   79

4. 29 avocados were in the basket. More avocados were added to the basket. Now there are 41 avocados. How many avocados were added to the basket?
   12

5. Steven has 32 peaches and Donald has 35 peaches. How many peaches do Steven and Donald have together?
   67

6. David has 66 apples and Brian has 15 apples. How many apples do David and Brian have together?
   81

7. Janet has 7 more bananas than Jennifer. Jennifer has 14 bananas. How many bananas does Janet have?
   21

8. 111 balls were in the basket. 71 are red and the rest are green. How many balls are green?
   40

Page 11

1. Some pears were in the basket. 10 pears were taken from the basket. Now there is 1 pear. How many pears were in the basket before some of the pears were taken?
   11

2. Amy has 33 fewer balls than Janet. Janet has 36 balls. How many balls does Amy have?
   3

3. 23 avocados are in the basket. 8 avocados are taken out of the basket. How many avocados are in the basket now?
   15

4. Brian has 6 apples. Steven has 25 apples. How many more apples does Steven have than Brian?
   19

5. 45 apricots are in the basket. 5 are red and the rest are green. How many apricots are green?
   40

6. 2 plums were in the basket. Some of the plums were removed from the basket. Now there is 1 plum. How many plums were removed from the basket?
   1

7. 33 peaches are in the basket. 9 are red and the rest are green. How many peaches are green?
   24

8. 19 bananas are in the basket. 19 bananas are taken out of the basket. How many bananas are in the basket now?
   0

Page 12

1. 49 marbles were in the basket. Some of the marbles were removed from the basket. Now there are 13 marbles. How many marbles were removed from the basket?
   36

2. 4 apples are in the basket. 4 apples are taken out of the basket. How many apples are in the basket now?
   0

3. 49 avocados are in the basket. 47 are red and the rest are green. How many avocados are green?
   2

4. Adam has 9 balls. Donald has 17 balls. How many more balls does Donald have than Adam?
   8

5. Sharon has 1 fewer pear than Jackie. Jackie has 13 pears. How many pears does Sharon have?
   12

6. Some peaches were in the basket. 13 peaches were taken from the basket. Now there are 12 peaches. How many peaches were in the basket before some of the peaches were taken?
   25

7. 10 apricots were in the basket. Some of the apricots were removed from the basket. Now there are 4 apricots. How many apricots were removed from the basket?
   6

8. Janet has 0 fewer bananas than Marin. Marin has 1 banana. How many bananas does Janet have?
   1

Page 13

1. 27 peaches are in the basket. 1 is red and the rest are green. How many peaches are green?
   26

2. Adam has 4 apples. Brian has 5 apples. How many more apples does Brian have than Adam?
   1

3. Marcie has 10 fewer marbles than Marin. Marin has 15 marbles. How many marbles does Marcie have?
   5

4. Some balls were in the basket. 19 balls were taken from the basket. Now there are 20 balls. How many balls were in the basket before some of the balls were taken?
   39

5. 39 plums are in the basket. 3 plums are taken out of the basket. How many plums are in the basket now?
   36

6. 28 bananas were in the basket. Some of the bananas were removed from the basket. Now there are 5 bananas. How many bananas were removed from the basket?
   23

7. 40 pears are in the basket. 20 are red and the rest are green. How many pears are green?
   20

8. 4 avocados were in the basket. Some of the avocados were removed from the basket. Now there are 3 avocados. How many avocados were removed from the basket?
   1

Page 14

1. 33 plums are in the basket. 1 is red and the rest are green. How many plums are green?
   32

2. Billy has 1 marble. Jake has 2 marbles. How many more marbles does Jake have than Billy?
   1

3. 25 avocados are in the basket. 11 avocados are taken out of the basket. How many avocados are in the basket now?
   14

4. Some peaches were in the basket. 5 peaches were taken from the basket. Now there is 1 peach. How many peaches were in the basket before some of the peaches were taken?
   6

5. 44 bananas were in the basket. Some of the bananas were removed from the basket. Now there are 31 bananas. How many bananas were removed from the basket?
   13

6. Janet has 0 fewer balls than Janet. Janet has 2 balls. How many balls does Janet have?
   2

7. 3 pears are in the basket. 2 are red and the rest are green. How many pears are green?
   1

8. Jennifer has 15 fewer oranges than Sandra. Sandra has 38 oranges. How many oranges does Jennifer have?
   23

Page 15

1. Jake has 18 apricots. Brian has 22 apricots. How many more apricots does Brian have than Jake?
   4

2. 8 avocados were in the basket. Some of the avocados were removed from the basket. Now there are 3 avocados. How many avocados were removed from the basket?
   5

3. Marcie has 15 fewer balls than Michele. Michele has 24 balls. How many balls does Marcie have?
   9

4. 19 bananas are in the basket. 18 bananas are taken out of the basket. How many bananas are in the basket now?
   1

5. 38 peaches are in the basket. 13 are red and the rest are green. How many peaches are green?
   25

6. Some marbles were in the basket. 11 marbles were taken from the basket. Now there are 12 marbles. How many marbles were in the basket before some of the marbles were taken?
   23

7. Some plums were in the basket. 9 plums were taken from the basket. Now there are 5 plums. How many plums were in the basket before some of the plums were taken?
   14

8. 3 pears are in the basket. 2 are red and the rest are green. How many pears are green?
   1

Page 16

1. Marin has 37 fewer marbles than Jennifer. Jennifer has 41 marbles. How many marbles does Marin have?
   4

2. 5 pears are in the basket. 3 are red and the rest are green. How many pears are green?
   2

3. 37 apples were in the basket. Some of the apples were removed from the basket. Now there are 21 apples. How many apples were removed from the basket?
   16

4. Some apricots were in the basket. 11 apricots were taken from the basket. Now there are 0 apricots. How many apricots were in the basket before some of the apricots were taken?
   11

5. 1 peach is in the basket. 1 peach is taken out of the basket. How many peaches are in the basket now?
   0

6. Billy has 1 banana. Adam has 28 bananas. How many more bananas does Adam have than Billy?
   27

7. 50 balls were in the basket. Some of the balls were removed from the basket. Now there are 36 balls. How many balls were removed from the basket?
   14

8. 42 avocados are in the basket. 37 avocados are taken out of the basket. How many avocados are in the basket now?
   5

Page 17

1. Michele has 1 fewer pear than Marin. Marin has 3 pears. How many pears does Michele have?
   2

2. Billy has 3 oranges. Steven has 25 oranges. How many more oranges does Steven have than Billy?
   22

3. 30 balls were in the basket. Some of the balls were removed from the basket. Now there are 0 balls. How many balls were removed from the basket?
   30

4. Some peaches were in the basket. 26 peaches were taken from the basket. Now there are 7 peaches. How many peaches were in the basket before some of the peaches were taken?
   33

5. 46 marbles are in the basket. 35 marbles are taken out of the basket. How many marbles are in the basket now?
   11

6. 13 apricots are in the basket. 10 are red and the rest are green. How many apricots are green?
   3

7. Michele has 25 fewer avocados than Marin. Marin has 30 avocados. How many avocados does Michele have?
   5

8. 30 apples are in the basket. 7 are red and the rest are green. How many apples are green?
   23

Page 18

1. Adam has 30 avocados. Brian has 34 avocados. How many more avocados does Brian have than Adam?
   4

2. 35 bananas are in the basket. 22 bananas are taken out of the basket. How many bananas are in the basket now?
   13

3. Some marbles were in the basket. 2 marbles were taken from the basket. Now there are 0 marbles. How many marbles were in the basket before some of the marbles were taken?
   2

4. 17 apricots were in the basket. Some of the apricots were removed from the basket. Now there are 0 apricots. How many apricots were removed from the basket?
   17

5. Michele has 1 fewer peach than Ellen. Ellen has 2 peaches. How many peaches does Michele have?
   1

6. 37 apples are in the basket. 20 are red and the rest are green. How many apples are green?
   17

7. 46 oranges are in the basket. 46 are red and the rest are green. How many oranges are green?
   0

8. 4 pears are in the basket. 3 pears are taken out of the basket. How many pears are in the basket now?
   1

Page 19

1. Some plums were in the basket. 23 plums were taken from the basket. Now there are 10 plums. How many plums were in the basket before some of the plums were taken?
   33

2. Marcie has 26 fewer marbles than Michele. Michele has 50 marbles. How many marbles does Marcie have?
   24

3. 5 apples are in the basket. 5 are red and the rest are green. How many apples are green?
   0

4. Brian has 25 balls. Jake has 39 balls. How many more balls does Jake have than Brian?
   14

5. 16 apricots are in the basket. 1 apricot is taken out of the basket. How many apricots are in the basket now?
   15

6. 44 pears were in the basket. Some of the pears were removed from the basket. Now there are 37 pears. How many pears were removed from the basket?
   7

7. 10 avocados are in the basket. 6 are red and the rest are green. How many avocados are green?
   4

8. Some oranges were in the basket. 7 oranges were taken from the basket. Now there are 5 oranges. How many oranges were in the basket before some of the oranges were taken?
   12

Page 20

1. 7 balls are in the basket. 1 is red and the rest are green. How many balls are green?
   6

2. Billy has 1 orange. Brian has 2 oranges. How many more oranges does Brian have than Billy?
   1

3. 35 apricots are in the basket. 20 apricots are taken out of the basket. How many apricots are in the basket now?
   15

4. Marcie has 4 fewer marbles than Marin. Marin has 15 marbles. How many marbles does Marcie have?
   11

5. Some bananas were in the basket. 22 bananas were taken from the basket. Now there are 6 bananas. How many bananas were in the basket before some of the bananas were taken?
   28

6. 35 pears were in the basket. Some of the pears were removed from the basket. Now there are 22 pears. How many pears were removed from the basket?
   13

7. 10 peaches are in the basket. 1 is red and the rest are green. How many peaches are green?
   9

8. 37 avocados were in the basket. Some of the avocados were removed from the basket. Now there are 34 avocados. How many avocados were removed from the basket?
   3

Page 21

1. Jackie's garden has 10 rows of pumpkins. Each row has nine pumpkins. How many pumpkins does Jackie have in all?
   90

2. Paul has two times more balls than Steven. Steven has five balls. How many balls does Paul have?
   10

3. Allan can cycle eight miles per hour. How far can Allan cycle in one hours?
   8

4. If there is one apricot in each box and there is one boxes, how many apricots are there in total?
   1

5. Audrey swims three laps every day. How many laps will Audrey swim in six days?
   18

6. David can cycle three miles per hour. How far can David cycle in two hours?
   6

7. Sharon's garden has 14 rows of pumpkins. Each row has 12 pumpkins. How many pumpkins does Sharon have in all?
   168

8. If there are 11 marbles in each box and there are two boxes, how many marbles are there in total?
   22

Page 22

1. Marcie has seven times more avocados than Adam. Adam has two avocados. How many avocados does Marcie have?
   14

2. If there are 13 bananas in each box and there are two boxes, how many bananas are there in total?
   26

3. Jake can cycle three miles per hour. How far can Jake cycle in 11 hours?
   33

4. Jake swims six laps every day. How many laps will Jake swim in 13 days?
   78

5. Michele's garden has nine rows of pumpkins. Each row has 11 pumpkins. How many pumpkins does Michele have in all?
   99

6. Audrey's garden has 14 rows of pumpkins. Each row has four pumpkins. How many pumpkins does Audrey have in all?
   56

7. Allan can cycle nine miles per hour. How far can Allan cycle in six hours?
   54

8. If there are two pears in each box and there are 13 boxes, how many pears are there in total?
   26

Page 23

1. Billy swims 12 laps every day. How many laps will Billy swim in 11 days?
   132

2. David can cycle seven miles per hour. How far can David cycle in six hours?
   42

3. Janet's garden has 12 rows of pumpkins. Each row has five pumpkins. How many pumpkins does Janet have in all?
   60

4. Paul has 15 times more plums than Janet. Janet has 15 plums. How many plums does Paul have?
   225

5. If there are four avocados in each box and there are six boxes, how many avocados are there in total?
   24

6. Steven can cycle nine miles per hour. How far can Steven cycle in nine hours?
   81

7. Brian has seven times more apricots than Jake. Jake has six apricots. How many apricots does Brian have?
   42

8. Audrey swims three laps every day. How many laps will Audrey swim in 15 days?
   45

Page 24

1. Amy's garden has three rows of pumpkins. Each row has five pumpkins. How many pumpkins does Amy have in all?
   15

2. Jake swims two laps every day. How many laps will Jake swim in 15 days?
   30

3. If there are four apples in each box and there are 13 boxes, how many apples are there in total?
   52

4. Allan can cycle seven miles per hour. How far can Allan cycle in seven hours?
   49

5. Janet has eight times more pears than Michele. Michele has nine pears. How many pears does Janet have?
   72

6. Jackie's garden has five rows of pumpkins. Each row has nine pumpkins. How many pumpkins does Jackie have in all?
   45

7. Adam can cycle two miles per hour. How far can Adam cycle in nine hours?
   18

8. Marcie swims 14 laps every day. How many laps will Marcie swim in nine days?
   126

Page 25

1. Brian can cycle 13 miles per hour. How far can Brian cycle in 13 hours?
   169

2. Audrey's garden has seven rows of pumpkins. Each row has 11 pumpkins. How many pumpkins does Audrey have in all?
   77

3. If there are three marbles in each box and there are nine boxes, how many marbles are there in total?
   27

4. Ellen swims six laps every day. How many laps will Ellen swim in three days?
   18

5. Sandra has one times more balls than Jackie. Jackie has 14 balls. How many balls does Sandra have?
   14

6. If there are 10 peaches in each box and there are eight boxes, how many peaches are there in total?
   80

7. Adam swims three laps every day. How many laps will Adam swim in seven days?
   21

8. Marin has 10 times more oranges than Allan. Allan has nine oranges. How many oranges does Marin have?
   90

Page 26

1. Billy can cycle 15 miles per hour. How far can Billy cycle in nine hours?
   135

2. Marcie's garden has 10 rows of pumpkins. Each row has one pumpkins. How many pumpkins does Marcie have in all?
   10

3. Jackie has three times more apples than Adam. Adam has one apple. How many apples does Jackie have?
   3

4. If there are six peaches in each box and there are 15 boxes, how many peaches are there in total?
   90

5. Audrey swims eight laps every day. How many laps will Audrey swim in eight days?
   64

6. Marcie has 14 times more balls than Janet. Janet has six balls. How many balls does Marcie have?
   84

7. Paul swims nine laps every day. How many laps will Paul swim in seven days?
   63

8. Janet's garden has 10 rows of pumpkins. Each row has four pumpkins. How many pumpkins does Janet have in all?
   40

Page 27

1. Billy swims 14 laps every day. How many laps will Billy swim in 10 days?
   140

2. If there are seven oranges in each box and there are five boxes, how many oranges are there in total?
   35

3. Sandra has two times more balls than Steven. Steven has 13 balls. How many balls does Sandra have?
   26

4. Janet's garden has two rows of pumpkins. Each row has nine pumpkins. How many pumpkins does Janet have in all?
   18

5. Brian can cycle 11 miles per hour. How far can Brian cycle in 11 hours?
   121

6. Jennifer has 10 times more pears than Sharon. Sharon has 13 pears. How many pears does Jennifer have?
   130

7. David swims two laps every day. How many laps will David swim in two days?
   4

8. Allan can cycle nine miles per hour. How far can Allan cycle in 12 hours?
   108

Page 28

1. Allan swims four laps every day. How many laps will Allan swim in 10 days?
   40

2. Amy's garden has two rows of pumpkins. Each row has six pumpkins. How many pumpkins does Amy have in all?
   12

3. Sharon has 12 times more oranges than Audrey. Audrey has 13 oranges. How many oranges does Sharon have?
   156

4. Paul can cycle one miles per hour. How far can Paul cycle in nine hours?
   9

5. If there are 14 marbles in each box and there are two boxes, how many marbles are there in total?
   28

6. David has one times more pears than Marin. Marin has 11 pears. How many pears does David have?
   11

7. If there are three apples in each box and there are seven boxes, how many apples are there in total?
   21

8. Sandra swims five laps every day. How many laps will Sandra swim in 12 days?
   60

Page 29

1. Jackie has 13 times more bananas than Janet. Janet has two bananas. How many bananas does Jackie have?
   26

2. Jackie's garden has 10 rows of pumpkins. Each row has 12 pumpkins. How many pumpkins does Jackie have in all?
   120

3. Donald swims 13 laps every day. How many laps will Donald swim in 12 days?
   156

4. David can cycle one miles per hour. How far can David cycle in 11 hours?
   11

5. If there are eight apricots in each box and there are 13 boxes, how many apricots are there in total?
   104

6. Allan can cycle 13 miles per hour. How far can Allan cycle in four hours?
   52

7. If there are six oranges in each box and there are four boxes, how many oranges are there in total?
   24

8. Brian swims six laps every day. How many laps will Brian swim in four days?
   24

Page 30

1. Allan has five times more apricots than Marin. Marin has seven apricots. How many apricots does Allan have?
   35

2. David can cycle three miles per hour. How far can David cycle in 11 hours?
   33

3. Marin's garden has one rows of pumpkins. Each row has three pumpkins. How many pumpkins does Marin have in all?
   3

4. If there are nine apples in each box and there are 14 boxes, how many apples are there in total?
   126

5. Jake swims 12 laps every day. How many laps will Jake swim in 15 days?
   180

6. Janet has 11 times more marbles than Marcie. Marcie has one marble. How many marbles does Janet have?
   11

7. If there are 12 oranges in each box and there are seven boxes, how many oranges are there in total?
   84

8. Sandra's garden has five rows of pumpkins. Each row has 11 pumpkins. How many pumpkins does Sandra have in all?
   55

Page 31

1. Billy ordered 12 pizzas. The bill for the pizzas came to $132. What was the cost of each pizza?
   11

2. Jake is reading a book with 57 pages. If Jake wants to read the same number of pages every day, how many pages would Jake have to read each day to finish in 19 days?
   3

3. A box of avocados weighs 266 pounds. If one avocados weighs 19 pounds, how many avocados are there in the box?
   14

4. How many 2 cm pieces of rope can you cut from a rope that is 22 cm long?
   11

5. You have 117 apricots and want to share them equally with 9 people. How many apricots would each person get?
   13

6. Marcie made 323 cookies for a bake sale. She put the cookies in bags, with 19 cookies in each bag. How many bags did she have for the bake sale?
   17

7. You have 104 apples and want to share them equally with 8 people. How many apples would each person get?
   13

8. Jackie made 36 cookies for a bake sale. She put the cookies in bags, with 2 cookies in each bag. How many bags did she have for the bake sale?
   18

Page 32

1. A box of apricots weighs 112 pounds. If one apricots weighs 7 pounds, how many apricots are there in the box?
   16

2. Brian is reading a book with 44 pages. If Brian wants to read the same number of pages every day, how many pages would Brian have to read each day to finish in 11 days?
   4

3. Marin made 36 cookies for a bake sale. She put the cookies in bags, with 2 cookies in each bag. How many bags did she have for the bake sale?
   18

4. You have 9 marbles and want to share them equally with 9 people. How many marbles would each person get?
   1

5. Steven ordered 3 pizzas. The bill for the pizzas came to $24. What was the cost of each pizza?
   8

6. How many 12 cm pieces of rope can you cut from a rope that is 156 cm long?
   13

7. Marcie made 90 cookies for a bake sale. She put the cookies in bags, with 9 cookies in each bag. How many bags did she have for the bake sale?
   10

8. You have 238 avocados and want to share them equally with 14 people. How many avocados would each person get?
   17

Page 33

1. Sandra made 28 cookies for a bake sale. She put the cookies in bags, with 7 cookies in each bag. How many bags did she have for the bake sale?
   4

2. How many 14 cm pieces of rope can you cut from a rope that is 14 cm long?
   1

3. Adam is reading a book with 176 pages. If Adam wants to read the same number of pages every day, how many pages would Adam have to read each day to finish in 16 days?
   11

4. Allan ordered 16 pizzas. The bill for the pizzas came to $304. What was the cost of each pizza?
   19

5. A box of peaches weighs 38 pounds. If one peaches weighs 2 pounds, how many peaches are there in the box?
   19

6. You have 42 plums and want to share them equally with 7 people. How many plums would each person get?
   6

7. Sharon made 190 cookies for a bake sale. She put the cookies in bags, with 19 cookies in each bag. How many bags did she have for the bake sale?
   10

8. A box of apricots weighs 20 pounds. If one apricots weighs 4 pounds, how many apricots are there in the box?
   5

Page 34

1. Donald ordered 16 pizzas. The bill for the pizzas came to $304. What was the cost of each pizza?
   19

2. How many 12 cm pieces of rope can you cut from a rope that is 144 cm long?
   12

3. Billy is reading a book with 88 pages. If Billy wants to read the same number of pages every day, how many pages would Billy have to read each day to finish in 11 days?
   8

4. You have 88 pears and want to share them equally with 8 people. How many pears would each person get?
   11

5. Amy made 95 cookies for a bake sale. She put the cookies in bags, with 5 cookies in each bag. How many bags did she have for the bake sale?
   19

6. A box of plums weighs 10 pounds. If one plums weighs 5 pounds, how many plums are there in the box?
   2

7. How many 3 cm pieces of rope can you cut from a rope that is 48 cm long?
   16

8. Billy is reading a book with 240 pages. If Billy wants to read the same number of pages every day, how many pages would Billy have to read each day to finish in 15 days?
   16

Page 35

1. You have 132 apples and want to share them equally with 12 people. How many apples would each person get?
   11

2. How many 12 cm pieces of rope can you cut from a rope that is 180 cm long?
   15

3. A box of apricots weighs 40 pounds. If one apricots weighs 20 pounds, how many apricots are there in the box?
   2

4. Sharon ordered 7 pizzas. The bill for the pizzas came to $56. What was the cost of each pizza?
   8

5. Jake is reading a book with 195 pages. If Jake wants to read the same number of pages every day, how many pages would Jake have to read each day to finish in 15 days?
   13

6. Amy made 48 cookies for a bake sale. She put the cookies in bags, with 8 cookies in each bag. How many bags did she have for the bake sale?
   6

7. Marcie ordered 12 pizzas. The bill for the pizzas came to $96. What was the cost of each pizza?
   8

8. A box of avocados weighs 266 pounds. If one avocados weighs 19 pounds, how many avocados are there in the box?
   14

Page 36

1. Jennifer made 24 cookies for a bake sale. She put the cookies in bags, with 4 cookies in each bag. How many bags did she have for the bake sale?
   6

2. How many 5 cm pieces of rope can you cut from a rope that is 95 cm long?
   19

3. You have 165 plums and want to share them equally with 15 people. How many plums would each person get?
   11

4. Billy is reading a book with 21 pages. If Billy wants to read the same number of pages every day, how many pages would Billy have to read each day to finish in 3 days?
   7

5. A box of oranges weighs 42 pounds. If one oranges weighs 3 pounds, how many oranges are there in the box?
   14

6. Audrey ordered 16 pizzas. The bill for the pizzas came to $112. What was the cost of each pizza?
   7

7. Allan is reading a book with 150 pages. If Allan wants to read the same number of pages every day, how many pages would Allan have to read each day to finish in 10 days?
   15

8. Allan ordered 15 pizzas. The bill for the pizzas came to $240. What was the cost of each pizza?
   16

Page 37

1. David ordered 9 pizzas. The bill for the pizzas came to $45. What was the cost of each pizza?
   5

2. Donald is reading a book with 30 pages. If Donald wants to read the same number of pages every day, how many pages would Donald have to read each day to finish in 2 days?
   15

3. A box of apricots weighs 32 pounds. If one apricots weighs 4 pounds, how many apricots are there in the box?
   8

4. You have 360 pears and want to share them equally with 18 people. How many pears would each person get?
   20

5. Amy made 126 cookies for a bake sale. She put the cookies in bags, with 14 cookies in each bag. How many bags did she have for the bake sale?
   9

6. How many 7 cm pieces of rope can you cut from a rope that is 35 cm long?
   5

7. Brian ordered 18 pizzas. The bill for the pizzas came to $288. What was the cost of each pizza?
   16

8. How many 4 cm pieces of rope can you cut from a rope that is 68 cm long?
   17

Page 38

1. You have 84 apricots and want to share them equally with 6 people. How many apricots would each person get?
   14

2. How many 14 cm pieces of rope can you cut from a rope that is 42 cm long?
   3

3. Sandra made 60 cookies for a bake sale. She put the cookies in bags, with 20 cookies in each bag. How many bags did she have for the bake sale?
   3

4. Jennifer ordered 9 pizzas. The bill for the pizzas came to $27. What was the cost of each pizza?
   3

5. A box of marbles weighs 20 pounds. If one marbles weighs 1 pounds, how many marbles are there in the box?
   20

6. Adam is reading a book with 11 pages. If Adam wants to read the same number of pages every day, how many pages would Adam have to read each day to finish in 1 days?
   11

7. Adam ordered 19 pizzas. The bill for the pizzas came to $19. What was the cost of each pizza?
   1

8. A box of bananas weighs 100 pounds. If one bananas weighs 5 pounds, how many bananas are there in the box?
   20

Page 39

1. Billy is reading a book with 160 pages. If Billy wants to read the same number of pages every day, how many pages would Billy have to read each day to finish in 8 days?
   20

2. Donald ordered 18 pizzas. The bill for the pizzas came to $144. What was the cost of each pizza?
   8

3. A box of apricots weighs 12 pounds. If one apricots weighs 4 pounds, how many apricots are there in the box?
   3

4. You have 136 marbles and want to share them equally with 8 people. How many marbles would each person get?
   17

5. Marcie made 64 cookies for a bake sale. She put the cookies in bags, with 8 cookies in each bag. How many bags did she have for the bake sale?
   8

6. How many 20 cm pieces of rope can you cut from a rope that is 20 cm long?
   1

7. A box of balls weighs 221 pounds. If one balls weighs 13 pounds, how many balls are there in the box?
   17

8. You have 200 pears and want to share them equally with 20 people. How many pears would each person get?
   10

Page 40

1. Janet ordered 12 pizzas. The bill for the pizzas came to $36. What was the cost of each pizza?
   3

2. How many 12 cm pieces of rope can you cut from a rope that is 156 cm long?
   13

3. Brian is reading a book with 8 pages. If Brian wants to read the same number of pages every day, how many pages would Brian have to read each day to finish in 1 days?
   8

4. You have 72 marbles and want to share them equally with 4 people. How many marbles would each person get?
   18

5. A box of pears weighs 180 pounds. If one pears weighs 10 pounds, how many pears are there in the box?
   18

6. Amy made 165 cookies for a bake sale. She put the cookies in bags, with 15 cookies in each bag. How many bags did she have for the bake sale?
   11

7. You have 40 plums and want to share them equally with 4 people. How many plums would each person get?
   10

8. Jackie made 24 cookies for a bake sale. She put the cookies in bags, with 6 cookies in each bag. How many bags did she have for the bake sale?
   4

Made in the USA
Middletown, DE
18 September 2022